Clima
~ loca

An enqui

elephant?

We a s growing in size and significa ce behind it, its likely conseque West Midlands have be e complex questions.

These ma many questions and d earing your thoughts.

Contents

Published by and available from:

Teachers in Development Education
Tide~ Centre
GO4 Millennium Point
Curzon Street
Birmingham B4 7XG

Tel 0121 202 3290
e-mail: wmc@tidec.org

ISBN: 0 948838 92 2

introduction

I don't believe in anthropogenic global warming in any way, and can prove it. I'm much more worried about the next ice age. We need the carbon dioxide for plants to live.

Professor David Bellamy
Warwickshire Education for Sustainable Development Conference, 24th May 2004.

The impacts of climate change will fall disproportionately upon developing countries and the poor.

Dr R K Pachauri, The Energy and Resources Institute, October 2004
[Source BBC News website http://newsvote.bbc.co.uk 20/10/04]

A challenge so far-reaching in its impact and irreversible in its destructive power that it alters radically human existence.

Tony Blair, British Prime Minister September 2004
[Source: Metro News 15/9/04]

In 1992, the Earth Summit at Rio de Janeiro set a process in motion which led to the Kyoto agreements on climate change. In 2005, these still awaited implementation, and were yet to draw in some of the world's biggest energy using nations.

In the meantime, "the approaching elephant" of climate change has been getting closer and bigger, and the urgency to respond at all levels – including education – has become more evident. These materials are a response from West Midland teachers who have been struggling with these dramatic and continuing debates.

The story of this resource

This particular story starts with a group of teachers, who visited The Gambia in 2002, and heard about the dramatic vulnerability of that West African country to rising sea levels, desertification and flooding. They heard from government agencies and others about what the effects of this might be on the country's water supply, fisheries, agriculture, natural environment and tourism industry, and how its capital city Banjul risked being lost within a few decades.

They also heard how countries like the UK produce a disproportionate amount of greenhouse gases such as CO_2, while low energy-using countries like The Gambia face many of the current and predicted consequences. As Dr Bernard Gomez, a scientist with the Water Resources Department told us, *"Your actions have an impact on the planet. What happens there can affect what happens here."* Gambian teachers, civil servants and environmentalists all stressed that education in the UK has a key role to play in addressing these inequalities.

The group returned to the UK and produced a discussion paper as a focus for teacher groups and workshops. Some initial ideas were published as part of the resource *Lessons in sustainability,* launched at a conference in Worcestershire in March 2003. At this event, teachers raised the need for an exploratory project about supporting children's enquiry into this complex issue.

This creative work began in Autumn 2003, through locally-based teacher groups in Birmingham, The Black Country, Coventry, Shropshire [with Telford] and Worcestershire. It drew on the expertise of group facilitators, climate change scientists, campaigners and local government officers. What is shared here has been refined from that process. It incorporates ideas and advice from other West Midland teachers, and insights from further study visits to The Gambia.

Thinking through climate change

Such a substantial and controversial issue requires a teaching approach which acknowledges the uncertainty behind our ideas about it: which invites children to decide for themselves about what is going on, and about appropriate action. These materials adopt a learner-centred enquiry approach, which may be adaptable for investigating other global issues.

All the schools used four key questions as a common framework:

1. What is climate change?

2. Why does it matter?

3. What can we do about it?

4. What have we learned and how?

The questions are based on Bloom's taxonomy [see box]. Page 8 shares more ideas about the value of taking an enquiry approach.

Our work offered opportunities for Primary, Secondary and Middle School children to learn to distinguish fact from opinion, and to think about the future. It offered space for them to determine their own thoughts and informed actions [often in real contexts, such as offering advice to the local authority]. It brought schools, age groups and departments together around a common focus.

Of course, curriculum enquiry is not the only possible educational response to climate change, which touches on many aspects of our schools and communities. In particular, there is a need [and an increasing expectation] for schools to think about institutional responses. For example:

- through energy or water efficiency, building and site design, transport policies, health issues like sunburn protection;
- offering spaces for children's views to be heard on issues like this, which affect their lives and engage them as citizens;
- developing schools' roles in the wider local and global community.

Page 10 shares some general questions and difficulties about sustainable development and climate change. The debates about this issue, and our role as educators, continue. We welcome your responses and creative ideas.

Thinking through climate change

Knowledge
– eg how is the climate changing?

Comprehension
– eg why is it changing? What is the greenhouse effect?

Application
– eg what is happening as a result? In what ways?

Analysis
– eg what are the causes and effects? Who wins and who loses?

Synthesis
– eg what can we do about it?

Evaluation
– eg are some solutions better than others? Which do I think are fairest?

Based on Bloom's taxonomy, cited in "Thinking through units", Fran Martin and Caroline Mathews, Primary Geographer 47, April 2002. See also "Thinking through the primary curriculum", Steve Higgins, Chris Kington Publishing, 2001.

West Midlands Coalition

This project contributes to the coalition by:

- exploring climate change issues, local and global;
- engaging teachers and partner organisations in developing ideas about how we enable children's enquiry into these issues;
- serving as a stimulus for further creative work on these themes, as part of Framework 2, *Sustainable development.*

See website **www.tidec.org**

Teaching about climate change: some issues

"This is something which is difficult to teach, it's okay to be unsure, nobody has all the answers ... I need your help with this."

Teaching about climate change raises a huge range of issues for us as teachers. It helps to discuss these with colleagues before starting work with children. The activity on these pages is adapted from a workshop, in February 2005, with a group of Herefordshire teachers, and makes use of items we found in the news.

On pages 6 and 7 we share some of the many educational issues, and suggested teaching strategies, which have come up in the course of our work, as a prompt to support your own discussions. You could cut statements out and jot down comments around each on a large sheet of paper to serve as a focus for discussion and consensus building. You will need to allow a good amount of time to do this.

Activity - looking at news stories

Ideas about climate change are altering all the time. We looked at some recent news stories about climate change, and considered:

- how do we react to these stories? What questions do they raise?
- what teaching issues come out of this?

We decided to choose some contrasting stories about climate change. Most newspapers carry regular stories, and we have found the internet a good source of news from different perspectives. Some responses are shared opposite.

Using the main headings on pages 6 and 7 as a prompt, we discussed the following points:

- there is a need for stories about uncertain science – all sorts of things were 'scientifically true' until proved otherwise, and those who challenged these 'truths' were not always popular at the time [eg Galileo, Darwin];
- we need to be able to help children take positive action, without giving a false impression – this might link in with initiatives such as Eco-Schools;
- what kind of knowledge base do teachers need? How much do we need to know to help prevent bias? When 'finding out together', how aware do we need to be of information sources at an appropriate level for children?
- we need to admit uncertainty to children, as a starting point for enquiry – *"This is something which is difficult to teach, it's okay to be unsure, nobody has all the answers ... I need your help with this."*

What points would you focus on?

Some useful sites for news: Transnational News Navigator [newspapers from around

Responding to news stories

Issues from story:

- Article focuses on Africa and impact on poorer communities
- Will a lack of 'voice' limit Africa's capacity to 'fight back'?

AFRICAN POOR TO BEAR BRUNT OF GLOBAL WARMING CRISIS

[Yahoo! news, 2/2/05]

Educational issues:

- Uncertainty – the story highlights that people are disputing climate science
- Are we 'happier' to talk about climate change impacts, rather than causes?

Issues from story:

- Conflict between what we say and what we do
- Are Russia and the USA out of Kyoto? What is happening?
- Oil industry – vested interests, better for USA

Blair enlisted to change the US climate

[Independent on Sunday, 20/2/05]

Educational issues:

- School computers on all weekend
- What can we do about it?
- Give positives – avoid doom and gloom
- Action is essential – and every little helps

Is Everest shrinking?

[Khaleej Times, United Arab Emirates, 24/2/05]

Educational issues:

- Needs a sea-change: more time devoted to these sustainability issues
- Difficult to do adequately in a cross-curricular way
- Significant knowledge base necessary for teacher to prevent bias
- Teaching children thinking skills, to debate evidence etc

Issues from story:

- Scaremongering?
- Confusing – is it getting smaller or not? What was the original height?
- How did Chinese measure it originally? What methods were used? How accurate were they?

Teaching issues

Planning around a live issue

1. You can plan via process. Providing you take an enquiry approach this issue is an ideal topic as it is always evolving and 'in the news'.

2. Encourage discussion, which brings in new ideas.

3. Be flexible: have a broad overview, but be prepared for tangents.

Ideas about teaching and learning

1. Investigating climate change challenges didactic approaches and promotes independent learning, as there are no clear answers. Build your professional confidence to open things up completely and let debate/ research flow.

2. Create opportunities for children to teach each other and adults. This can be very appealing. They need to know that we can also learn from them. This can also support children who lack confidence in challenging authority.

3. Consider what level of support children need. To what extent should the enquiry process be controlled/open? Do you need an end-point as a focus and support for children?

4. Plan for open-ended outcomes [this can be hard for children]. Decision-making activities can help keep minds open.

The gap between what the school models and what children learn

1. Involve the whole school in promoting a sustainability ethos. Give children opportunities to make a difference [eg through conducting school audits, creating action plans].

2. Encourage all people in the school to work on a given set of standards [eg closing doors, turning lights off], which encourage children to see that this is not just a separate lesson but a way of life.

3. Consider a whole-school [or cluster] approach to enquiry learning, following work on climate change, and rolled out across subject areas and year groups.

Responding to the human dimensions of climate change

1. Address emotional responses as they arise: provide opportunities for children to discuss, articulate feelings, acknowledge and share thoughts, offer positive responses.

2. Handle discussions sensitively. Children can get very distressed about what they see or experience. Create spaces to find out about children [eg some children's lives may have been touched by events which they connect to climate change].

3. Empowerment. Give a balanced, hopeful view on what individuals and groups can do to achieve sustainability: they can improve things. Impart a sense of shared responsibility.

Being both positively-focused yet realistic

1. Share the idea that many people [eg governments] are taking this seriously but no-one has the answers yet. Therefore, what the children come up with is very important.

2. Stress that every little helps: ant-power!

3. Be positive: consider solutions, give examples, share success stories. Changes that are 'good' for climate change are often 'good' for the planet generally.

"No-one has the answers yet. Therefore, what the children come up with is very important."

Bias and controversy

1. Use activities which sort 'facts' from opinion, certainty from uncertainty, explore ideas about bias and controversy.

2. Be open about bias being present as a chance to develop critical thinking. Offer a forum for discussion and enquiry, to help 'balance' views and access varied perspectives.

3. Role play – arguing from different perspectives, questioning motives, leading to informed persuasive argument.

Citizenship and positive responsibility

1. Explore what is being done in the local area, and what should be happening. Avoid blame: emphasise individual, collective and international action. Everyone has a part to play. What can we do as an individual and a school?

2. Share the idea that doing something very small can make a difference [eg energy saving, recycling, water conservation].

3. Invite children to consider their responsibilities. How do their actions affect others? What actions have an effect on them? Widen the issue out to a global scale.

Teaching in the context of uncertainty

1. Be honest and open with the children. Make them aware of the uncertainties; nobody knows the future.

2. Focus on a range of opinions from a variety of websites and other sources, including news media.

3. Explore children's own ideas and share some current thinking – encourage debate.

Teaching about a complex issue

1. Admit to not 'knowing everything' – exploring with the children is an essential starting point.

2. Enquiry based learning, being child-centred, means learners should move forward within their own understanding. They will set their own levels of complexity.

3. Use accessible sources and materials which give children the freedom to start at their own level [eg news articles].

Helping children make links to the 'big picture'

1. To emphasise contrasts, start with a local/personal issue, then broaden out [eg floods in Ironbridge, talking with grandparents about change, to global patterns of flooding].

2. For a sense of context and systems, start with the 'big picture' and then focus on specifics [eg looking at sea level rise worldwide and then what is happening at the coast – in Bangladesh and East Anglia].

3. Build empathy by offering individual stories or images which children can remember, relate to and understand.

Why take an enquiry approach to teaching about climate change?

A group of us brainstormed these lists, having supported some enquiry into climate change in our own schools. One lists arguments for taking an enquiry approach. The other raises some potential difficulties. You may want to add your own statements, amend these lists, or do your own lists and compare them.

Reasons for enquiry

It involves transferable knowledge, skills and understandings	It allows children to explore a variety of perspectives	Children are able to challenge their own preconceived ideas [and each other's ... and ours]
It encourages cross-curricular opportunities	There are no 'right or wrong' answers to climate change issues	It allows children to explore, affirm and develop their values
Ofsted and QCA encourage student-led enquiry learning	It's a real issue, it's live, it matters	It encourages ownership of the debate [and therefore of informed choices which arise from it]
It empowers learners and helps them believe they might do something about it	It's a complex and controversial issue – and enquiry helps avoid us enforcing our own bias	It stretches our own teaching towards a more enabling style
It empowers children to take control of their own learning	It encourages collaborative work	Children's opinions matter
Motivational	Encourages creativity	Encourages children to explore a wide range of resources
Children feel they can make a difference	Encourages higher-order, independent learning [including learning about learning]	It's a complex and changing issue – this approach reflects its reality
Develops children's thinking skills	Add your own ...	Add your own ...

Useful sites on enquiry learning: www.unesco.org/education/tlsf [unit 21]

Some difficulties about taking an enquiry approach

May challenge some children – due to preferred learning styles or the intimidating level of the task and nature of subject	Time constraints
Resource constraints eg ICT	Classroom management
In tension with ideas about curriculum coverage	Issues about formal outcomes – recording the process adequately
Children may come up with views that we object to	Debate about clear end-points versus open-ended enquiry – some children need more definition than others
Need to pay attention to grouping and groupwork skills	Requires a climate where children are used to listening and speaking
May be difficult to recognise outcomes	May 'leave it hanging' too much
Teaching relationships may be challenged [eg teacher as 'expert']	Requires teacher confidence [eg in own subject knowledge, but also in letting go]
It may show up disparities between what is asked of children and the way schools behave	It is not the only valid teaching approach possible

Activity

for a staff or departmental meeting, or INSET session.

- We cut these statements up and spread them around.
- Each teacher chose one priority – which they shared with a colleague or small group. Why did they choose this one?
- Alternatively, you could highlight your first three priorities.

In groups we discussed:

- given our priorities, what did this mean for the school's enquiry into climate change? [eg the objectives we set or the kind of outcomes we might be looking for]
- did it raise particular challenges or opportunities for the school?
- was there additional support which might be needed?
- were there institutional implications for the whole school?
- what were the main difficulties [see above]?
- how would we address them?

Sustainable development and climate change - is it going to rain tomorrow?

Sustainable development means many things to different people – over 200 definitions have been published. It is a continuing process, which we access through our own informed engagement, and therefore it contains a crucial element of uncertainty. There are no clear answers, and this is reflected in the climate change debate.

Nonetheless, there is substantive content to be explored, and an urgent need for action at a variety of scales. Learning is an essential part of sustainable development: to bring about change, and to explore solutions creatively.

You could use the following quotations from around the world to explore what sustainable development means to you [and the children or young people you teach]. There is no single right or wrong definition!

Key for images [from top left]

Mural in Gambian school;
Eskimo children, Alaska*;
Waste dumped in UK forest⁺;
Flooding in Carlisle*;
Wind farm in UK⁺;
Gambian Renewable Energy Centre;
Alaskan forest fire*;
Solar box cooker, The Gambia;
Coal-fired power station, UK⁺;
Coastal erosion in The Gambia;
Elan Valley reservoir, Wales⁺;
Gambian mangroves.

*Photographer: Ashley Cooper.
All other images: Tide~ [⁺from *Learning today with tomorrow in mind*]

"REP 123, Resources, Environment, People living in harmony!"

Ruth Henshaw, [Shropshire teacher, UK, 2004]

"I believe there has to be a happy lifestyle medium somewhere between hair shirt hippy and toxic Texan"

Tamasin Cave, Former Editor, Ergo Magazine [London, UK, 2004]

"Sustainable development: You have to put back. It's like anything in life – if you just keep taking then eventually there will be nothing left"

Garry Greenland, organic supplier [UK, 2001]

"It's all about caring for things so they can live"

5 year old on outdoor programme [Worcestershire, UK, 2004]

"Sustainable development: development that meets the needs of the present without compromising the ability of future generations to meet their own needs"

Gro Harlem Brundtland [Norway, 1987]

"One cannot live for half the day concerned with the environment and the other half ignoring or destroying it"

Suryo Prawiroatmodja [Indonesia]

"A global human society, characterised by islands of wealth, surrounded by a sea of poverty, is unsustainable"

Thabo Mbeki, [South Africa, 2002]

"We are the guests, not the masters, of nature and must develop a new paradigm for development and conflict resolution, based on the costs and benefits to all peoples and bound by the limits of nature herself rather than by the limits of technology and consumerism"

Mikhail Gorbachev [Russia, 2005]

"All of life is interrelated. We are all caught in an inescapable network of mutuality, tied to a single garment of destiny. Whatever affects one directly affects all indirectly "

Martin Luther King [USA, 1965]

Developing an understanding of climate change is part of educating for sustainable development. It includes: the control of greenhouse gases, land and energy use, pollution and transport, desertification, rising sea levels, effects on biodiversity [changes in biological populations and distribution, species extinction], changing weather patterns, climate change refugees.

Some useful websites - Carbon Trust www.thecarbontrust.co.uk Intergovernmental

REP 123: A framework for exploring sustainable development

Exploring this 'big' issue is about asking questions and exploring solutions. We offer a number of images relating to the REP 123 framework: Resources, Environment, People. What climate change questions do these images provoke for you? [We offer some prompt questions to get you going]. How are the questions in the three areas connected?

"The world is never going to run out of energy ... the question is how we move from one means of generating energy to another"

Sir Crispin Tickell, 2005

"We can transform our denial of climate change into positive action, if we have the will"

Rev John Oliver 2005

Panel on Climate Change www.ipcc.ch   BBC Weather Centre www.bbc.co.uk/climate

Planning an enquiry on climate change - curriculum links

Our work has focused on cross-curricular enquiry skills. This has often meant bringing the teaching of different subjects and age groups together – sometimes across schools and key stages – and we have seen that as a strength and opportunity.

It has also referred to frameworks for teaching and learning about sustainable development in the curriculum. This includes the seven key concepts for ESD, which form the core of QCA's guidance.

Such concerns inevitably go beyond the formal curriculum, into the life and ethos of the school, and its relationship with wider communities at all scales. It has helped give our schools and clusters a sense of purpose as 'learning communities' engaging with real issues which touch on our lives. To support these, we have made links to initiatives such as Eco-Schools and Healthy Schools Initiative, School Travel Plans and schemes tying school meals to local food production. Many of us have been inspired to undertake a sustainability audit of the school, using tools such as Thinkleadership [below].

We offer here a grid showing links between an enquiry into climate change, National Curriculum Programmes of Study and some key QCA Units of work. This is by no means a definitive table, but it may help you see that it is possible to undertake such an enquiry as a means of fulfilling statutory requirements in a way which engages learners in a meaningful, purposeful and motivational way.

Key concepts from the Holland Report on education for sustainable development

Examples of starting points

Years 3 & 4	Stirchley Primary, Shropshire *Literacy:* key vocabulary, developing discussion skills, creating 'talk stories'. *Art and design:* poster and badge making. *Citizenship/Geography [Link to Eco-Schools]:* monitoring school energy use, thinking skills, 'mysteries'.
Year 4	Somerville Primary, Birmingham *Geographical enquiry/Citizenship:* Creating concept map of responsibilities for informed action; special week of activities with outside agency [Severn Trent Water].
Years 5 & 6	Bredenbury Primary, Herefordshire *Literacy:* speaking and listening skills, participation in debate, linking Hindu storytelling ['every little helps'] to climate change, exploring fact and opinion in mixed ability groups. *Geography, linked to ICT*: 'In the news' – using secondary sources to gather information about climate change, looking at fact and opinion.
Years 6 & 7	Kingswinford Secondary and St Mary's CE, Dudley *Citizenship* as lead subject in a bridging unit, involving peer learning across schools and using a summer heatwave as a starting point.
Years 7-9	Wakeman School and The Burton Borough School, Shropshire *Geography:* web enquiries; comparing 'Mediterranean' and 'Big Chill' scenarios; considering core concepts and possible solutions.

Whole school responses - Thinkleadership www.thinkleadership.org.uk

	Literacy	Science	Geography	DT and ICT	Ctizenship/PSHE	Other
Enquiry	News reports, journalism	Sc1 Enquiry: ideas & evidence; investigative skills; using secondary sources; considering evidence KS3 Unit 9M Investigating scientific questions Unit 5/6H Enquiry in environmental contexts	Enquiry & skills: investigating, evaluating evidence and presenting KS2 Unit 16 What's in the news?	KS2 ICT Unit 6D internet searches and interpretation KS3 ICT Unit 10 Information – reliability, validity and bias Unit 14 Global communication	Knowledge & understanding – becoming informed citizens KS2 Citizenship Unit 5 Living in a diverse world Unit 7 Human Rights KS3 Citizenship Unit 10 Debating a global issue Unit 21 People & the environment	KS3 RE Unit 7E What are we doing to the environment? Maths – Surveys, graphs and data handling; calculating carbon emissions etc; probability
Participation	Persuasive writing, discussion texts, debates: argue point of view Drama Balance & evaluate different points of view		Knowledge & understanding of places [interdependence, global citizenship] KS2 Units 13/22 Contrasting localities	KS2 Unit 5F Monitoring environmental changes	Skills of participation and action KS2 Citizenship Unit 1 Taking part Unit 11 In the media PSHE 2a Research, discuss & debate topical issues	KS2 RE Creation stories [especially non-European traditions]
Processes		Sc2 Living things Sc3 Materials KS2 Unit 4B Habitats Unit 5C Gases around us Unit 6A Interdependence & adaptation KS3 Unit 9G [8]: Is global warming happening?	Knowledge & understanding of patterns & processes [impacts on places and environments] KS2 Unit 10 Weather patterns over Europe Unit 11 Water Unit 17 Global Eye	KS3 Design technology: renewable materials, technical solutions, evaluating processes & products KS2 ICT Unit 5C Evaluating information	Skills of enquiry & communication; analysing information & sources; justify a personal opinion; contribute to group discussions, take part in debates KS2 Citizenship Unit 1 Taking part Unit 2 Choices	Art & design 2a Investigate, combine & manipulate materials & images, taking account of purpose & audience
Choices	Non -chronological reports, explanations	KS3 Sc4 Physical processes Energy resources & energy transfer Benefits & drawbacks of scientific developments [eg for environment & quality of life] Unit 7I Energy resources Unit 9I Energy & electricity	Knowledge & understanding of environmental change & sustainable development KS2 Unit 21 Improving the area KS3 Unit 4 Flood disaster – how do people cope?	KS2 ICT Creating multimedia and Powerpoint presentations KS3 Presentation	KS2 PSHE 2j/f Resources can be allocated in different ways 5d Choices & decisions Citizenship Unit 2 Choices Unit 7 Children's Rights	KS2 History Units 7/8 The Victorians: Industrial revolution & use of resources KS3 History – oral & community History, change over time

1 What is climate change?

Starting points

Many of us started our enquiries with something which allowed children to share some of their existing ideas and understandings about this question [eg as a mind map]. We then returned to this at the end of the whole enquiry [see page 26].

We found it helped for children to come up with their own further questions in relation to all four of the key questions [some prompts are offered, right].

We introduced the whole enquiry cycle at this stage, and encouraged children to review what and how they were learning as they went along [eg by putting a 'map' of the cycle on the wall, and adding to it as work proceeded – see page 26].

Some children produced questionnaires for parents, asking what they knew about climate change. Others interviewed grandparents – have things changed? How?

Activities for all four key questions offered opportunities for children to feed back to peers, through news reports, PowerPoint presentations, assemblies, leaflets, video and so on.

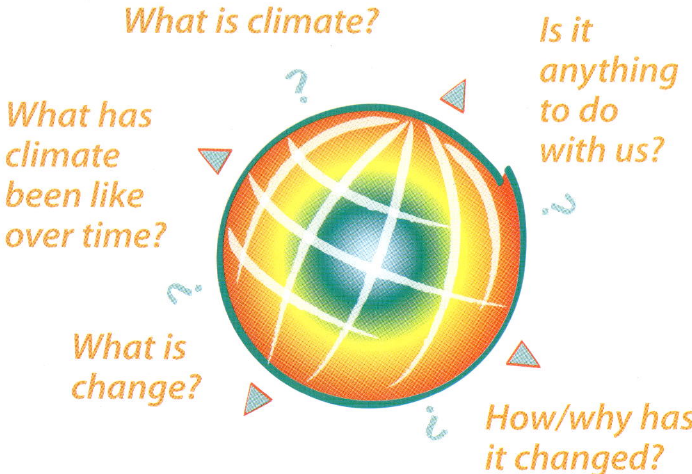

Key words

Children explored key climate change concepts and terms. The cards opposite were used at both Key Stages 2 and 3 as a stimulus for initial discussions.

- We organised the class into groups.
- Each group was given a picture card and a set of suggested words.
- They were given time to devise sentences containing the given words in relation to the picture.
- Each group fed back to the others, explaining their picture using the words.
- To extend this activity, you could use the definitions below, and children could discuss how similar or different they are to their own ideas.

DEFINITIONS

Climate change	**Global warming**
The Earth has always got hotter and cooler, wetter and drier, over long periods of time. These changes are part of natural cycles, but changes are also affected by what people do. Some scientists are still debating how much this is the case.	The Earth has become hotter in the last 100 years. Sea levels are rising, ice sheets melting, deserts spreading.
Greenhouse effect	**Carbon sink**
Gases in the Earth's atmosphere trap heat and reflect it back to the surface. This helps stop us from freezing. Too much, however, may cause overheating. Some of these gases [like methane and carbon dioxide] are released through what people do eg burning oil and coal.	All plants absorb carbon dioxide, and give out oxygen into the atmosphere. Large forests and ocean plant life are important for keeping the planet healthy. Carbon is an essential part of all living things.

Child-friendly sites - DEFRA: www.defra.gov.uk/environment/climatechange/schools
CBBC - search 'climate change' for up-to-date information: www.bbc.co.uk/cbbc

Climate change

wetter
planet earth
time
natural cycle
hotter
climate
cooler
drier

Global warming

ice sheets melting
carbon dioxide
flooding
hotter
planet earth
last 100 years
rising sea levels
deserts spreading

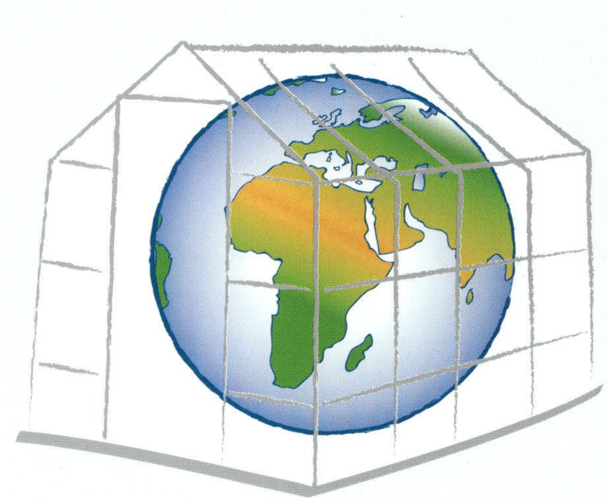

Greenhouse effect

overheating
heat trapped
increased gases
atmosphere
reflected heat
methane
carbon dioxide
planet earth

Carbon sinks

carbon
carbon dioxide
atmosphere
oxygen
planet earth
absorb
healthy balance
plants

What is climate change? www.climatechange.gc.ca/english/climate_change
Phenology - investigate changing seasons: www.woodland-trust.org.uk/phenology

1. What is climate change?

Moving images

It was important early on for us to help children put climate change into context, to start to develop empathy and to stimulate thinking/discussion. These images can be used to support these aims.

- We organised the class into pairs or groups.
- Each group was given an image.
- Students had to imagine what went before, and the after effects of, the situation/event. What might the people be feeling? What were the possible causes of this situation/event? We offered stimulus questions – *what? where? when? who? why?*
- Younger children drew a picture of themselves on post-it notes, and placed that on the picture. They talked about what they could see, hear, feel, smell, touch 'in the picture'.
- Children were hot seated in a plenary, voicing the thoughts and feelings of people in the images.

Fact or opinion?

Several of us made use of QCA KS2 Geography Unit 16 *"What's in the news?"*

- Children made a start by doing an internet search on 'climate change'.
- They copied and pasted statements onto two documents: 'fact' and 'opinion'.
- We limited this to one sheet of A4 [on the computer, or as an actual sheet].
- Work was [printed and] displayed to stimulate further discussion.
- We offer some useful and child-friendly websites on pages 14 and 15.
- As a plenary, children worked in groups. Each group chose a statement about climate change to read out to the class.
- The other groups had two cards – 'fact' and 'opinion', and had to decide which to use for the statement being read out, and explain the reasons for their decision.
- Call My Bluff – older children introduced some 'red herring' statements, and invited the rest of the class to identify which were true/real statements and which were 'bluffs'.

Looking at websites in this way proved a good way of linking up Science and Geography departments in Secondary schools.

Talking information

The graph below comes from the Intergovernmental Panel on Climate Change [IPCC], and shows temperature variations over the last 140 years.

- Children worked in pairs or small groups.
- They were asked to look at the graph carefully, and decide on two or three points over that period where climate had changed significantly [it may have got hotter or cooler].
- They wrote speech bubbles for people living at those times, saying something which related to the change.
- It helped to offer children roles when they made their statements [eg what would they say if they were a farmer? a holidaymaker? if they worked in a ski resort? if they lived in Alaska? Algeria? Bangladesh?].

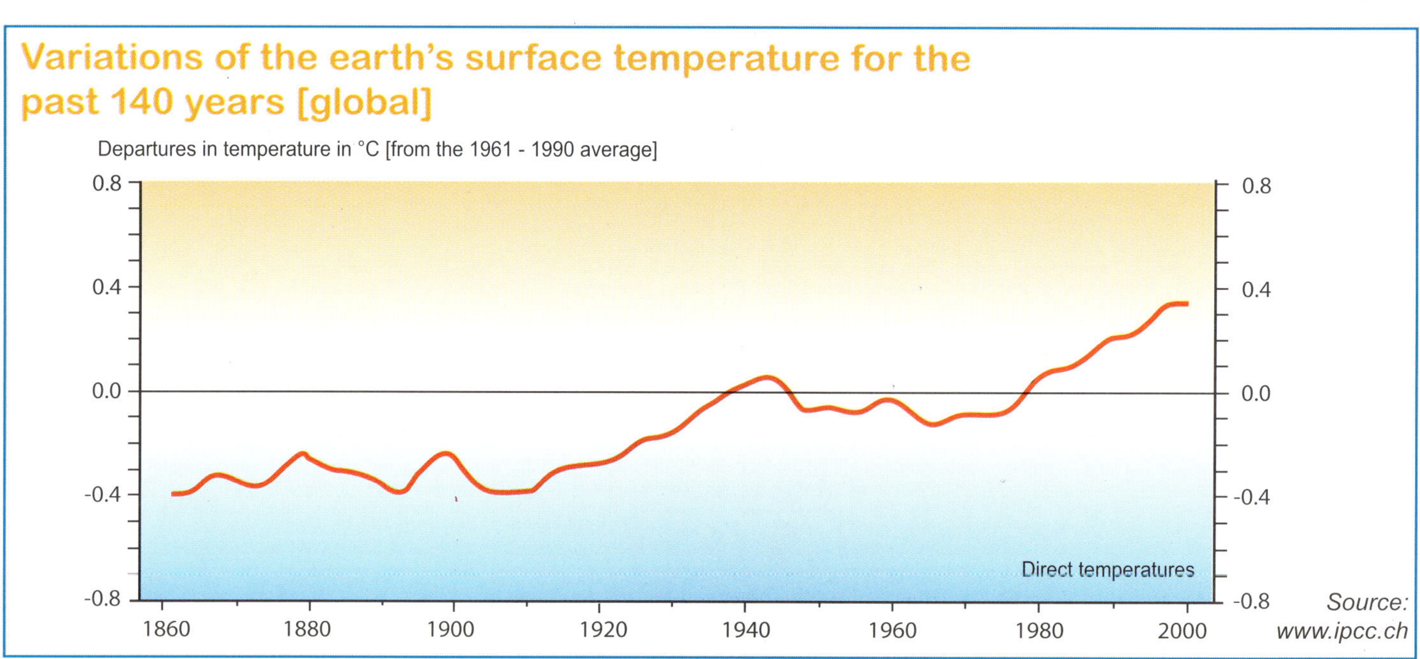

Images: Earth from the air: www.yannarthusbertrand.com
Climate change data and graphs: IPCC site: www.ipcc.ch

Moving images

Photos from top:

- House collapsed in Alaska due to melting permafrost
- Residents being boated to safety during Carlisle floods, 2004
- Warning sign at glacier which is rapidly retreating. Kenai Fjords National Park, Alaska

[Photographs: Ashley Cooper]

Search 'climate change' + image on www.google.com
Animated climate change timeline: www.cru.uea.ac.uk/cru/demos/temrec/nh.htm

2 Why does it matter?

In the news

We used local news reports [eg floods, building on floodplains, hosepipe bans] as a way in to this question. Children took on the role of a newscaster, investigating and reporting on the effects of climate change in the UK. Their brief was to include something about their own area, and two contrasting localities [eg coastal erosion, flooding, drought, tourism, farming]. Not all of these were necessarily negative [eg growing sunflowers and grapes].

Children grouped headlines and quotations around key ideas: climate change, business, the environment, people, costs, positives, negatives. They looked at how apparently negative consequences could become opportunities. [eg Holland, with a history of flooding, now exports expertise to places starting to face similar challenges].

Why does climate change matter to me?

The images opposite are of the same local area in 2030. Alternatively, children could look at a picture of their local area and draw their own images.

- ❑ Children looked at the images and considered what was in them, using the prompt questions opposite.
- ❑ They made choices: which picture would I choose to live in and why?
- ❑ We explained that these images show four different ways that the climate might change.
- ❑ Children looked at their choices again: had they reconsidered anything? Why?

Some of us linked this activity to work on oral history [eg by collecting quotations] or old maps/photos [eg tithe maps] to explore how people have influenced the environment around the local area. Local record offices and newspapers have these materials. Having considered these, children made predictions for the local area in future.

The day after tomorrow, tomorrow, tomorrow ...

The wheel opposite shows several scenarios for climate change. Children worked in groups, using the wheel as a proforma, discussing and recording the consequences of each change. You might want to prompt children about particular issues: health, farming, wildlife, leisure, housing etc.

Alternatively, you could copy it onto card, and put a cocktail stick through the middle. Children could then spin it, and consider the scenario which it lands on.

This activity was inspired by an idea in *Teaching about climate change* by Tim Grant and Gail Littlejohn [for details of this resource see page 30]

Children from Kingswinford School in Dudley used a summer heat wave as the starting point for their investigation. As a stimulus, they used material downloaded from www.upd8.org.uk Source: Centre for Science Education and ASE

Local newspaper websites [eg www.thisisherefordshire.co.uk] often have archives of past stories on flooding etc.

Climate change scenarios ~ 2030

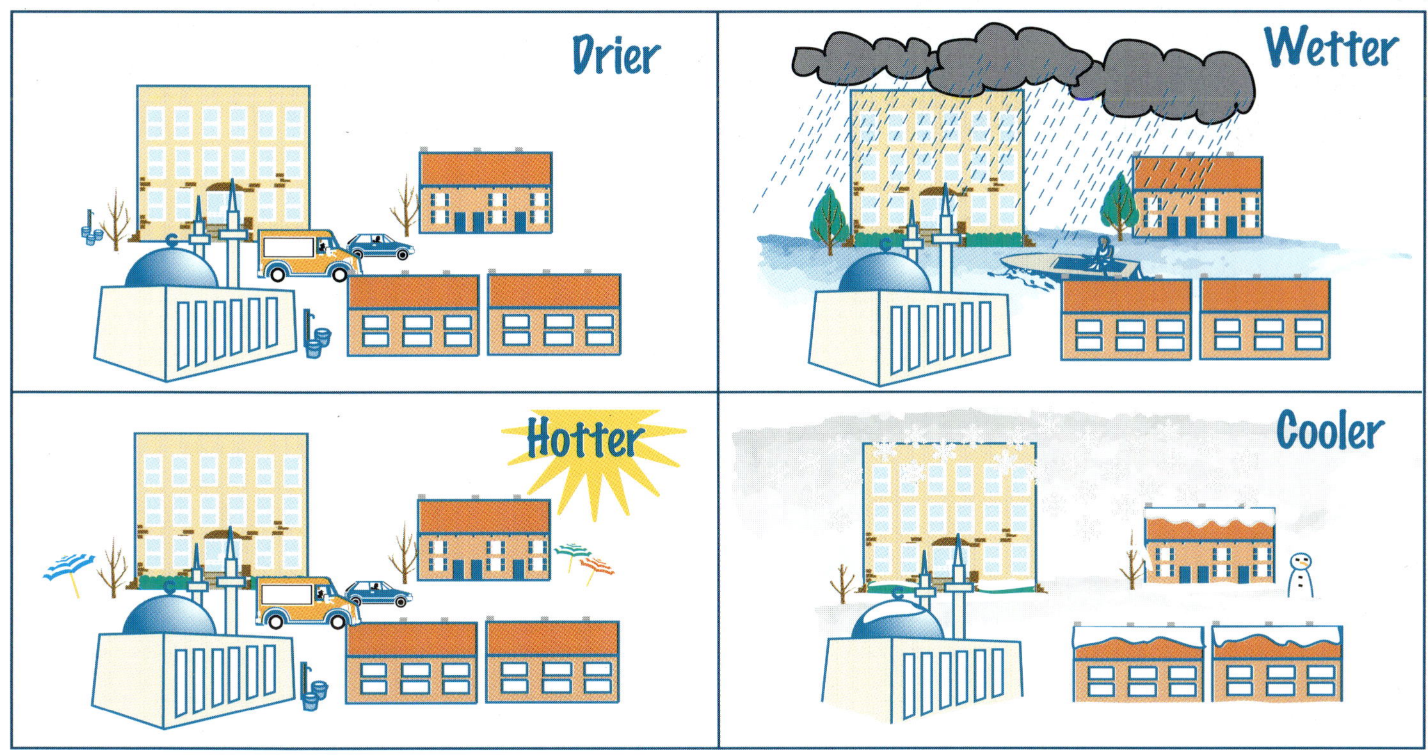

Climate change wheel

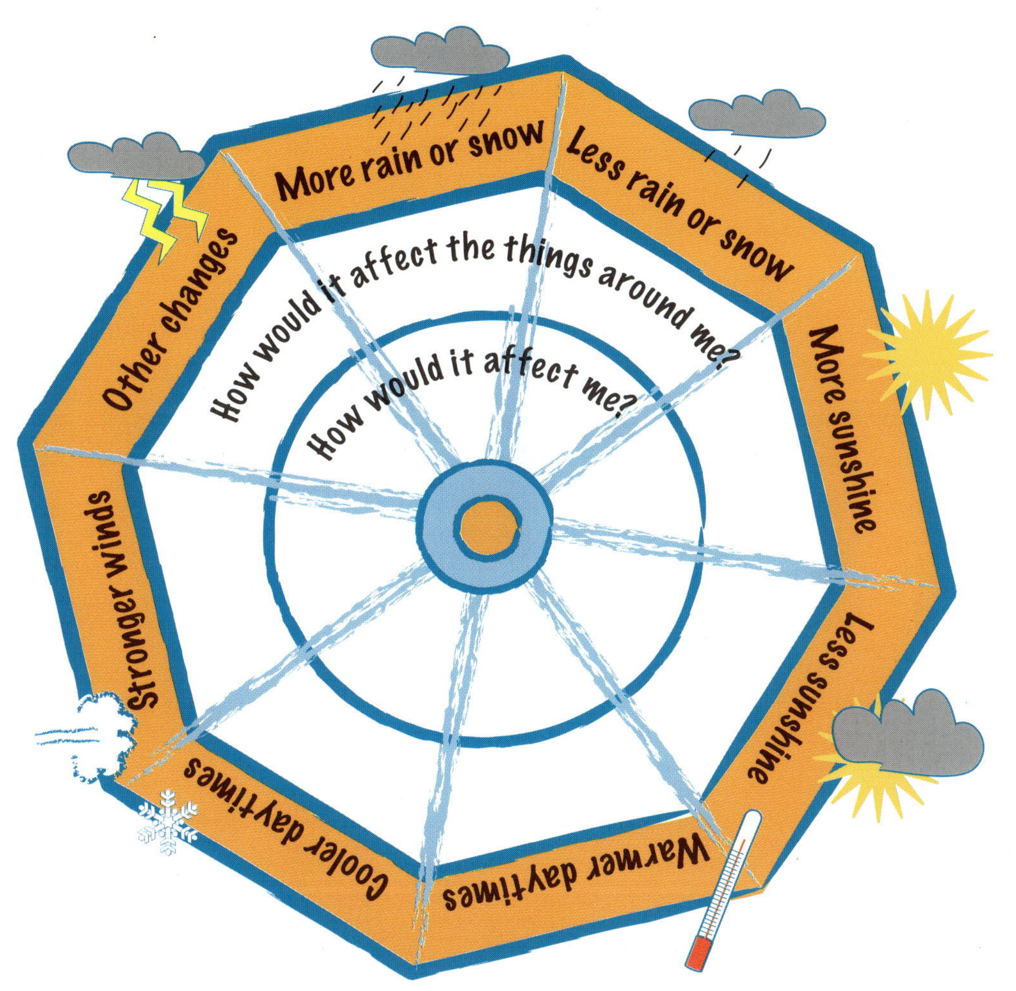

What can I see?

What might I hear?

What might I smell?

How would I feel living there?

Which picture would I choose to live in and why?

TNN links to papers from all over the world www.transnational.org/new/TNN.html
For the hottest and coldest places forecasted worldwide www.bbc.co.uk/climate/

Why does it matter? ... globally

"Discussion is sometimes the most important thing"

Considering change

Year 3 children at Stirchley Primary, Telford, considered what change means.

- ❑ They looked at a range of images from around the world, including art/drawn images, as well as photos. These all dealt with change in one form or another. The images also served as a stimulus for evocative language.
- ❑ Children tried to identify what linked the images.
- ❑ They discussed the idea that change, including climate change, is occurring at a variety of scales, including nationally and globally.
- ❑ They were asked, *"Why does change matter?"* [Especially if there are positive as well as negative factors].
- ❑ Some sources of global images are listed below.

Mysteries: linking the local to the global

To bring global connections to the fore, teachers wrote 'mysteries' for children to unravel. These helped children see how an individual's action here can affect another's life on a global scale, and so were linked to local contexts.

They developed their own mysteries, linked to their own contexts.

- ❑ Our example opposite is from Stanley Road Primary in Worcester, which had experienced recent floods. Years 3 and 4 children worked in two groups, looking at Worcester and Bangladesh. Groups then compared notes, and linked the two stories together.
- ❑ Many children at Somerville Primary School in Birmingham are of South Asian heritage. Their mystery linked a daily journey to school by car to a Bangladeshi market trader going out of business.
- ❑ Bredenbury Primary in rural Herefordshire linked the secretary's car journey to school to effects on a community in The Gambia.

Children could follow this activity up by hot-seating the main characters in their mystery.

Ideas adapted from Mysteries make you think *[Leat & Nichols, Geographical Association, 1999]*

Talking graphs

Children looked at a blank graph, such as the one below on energy use worldwide. They worked together to interpret the graph, and then looked at it from the point of view of people living in different places. For example, what might someone in Africa ask of us in Europe?

"Africa as a whole contributes about 3% of global greenhouse gas emissions"

Bubu Jallow, Director of Water Resources, The Gambia

Average energy use per person

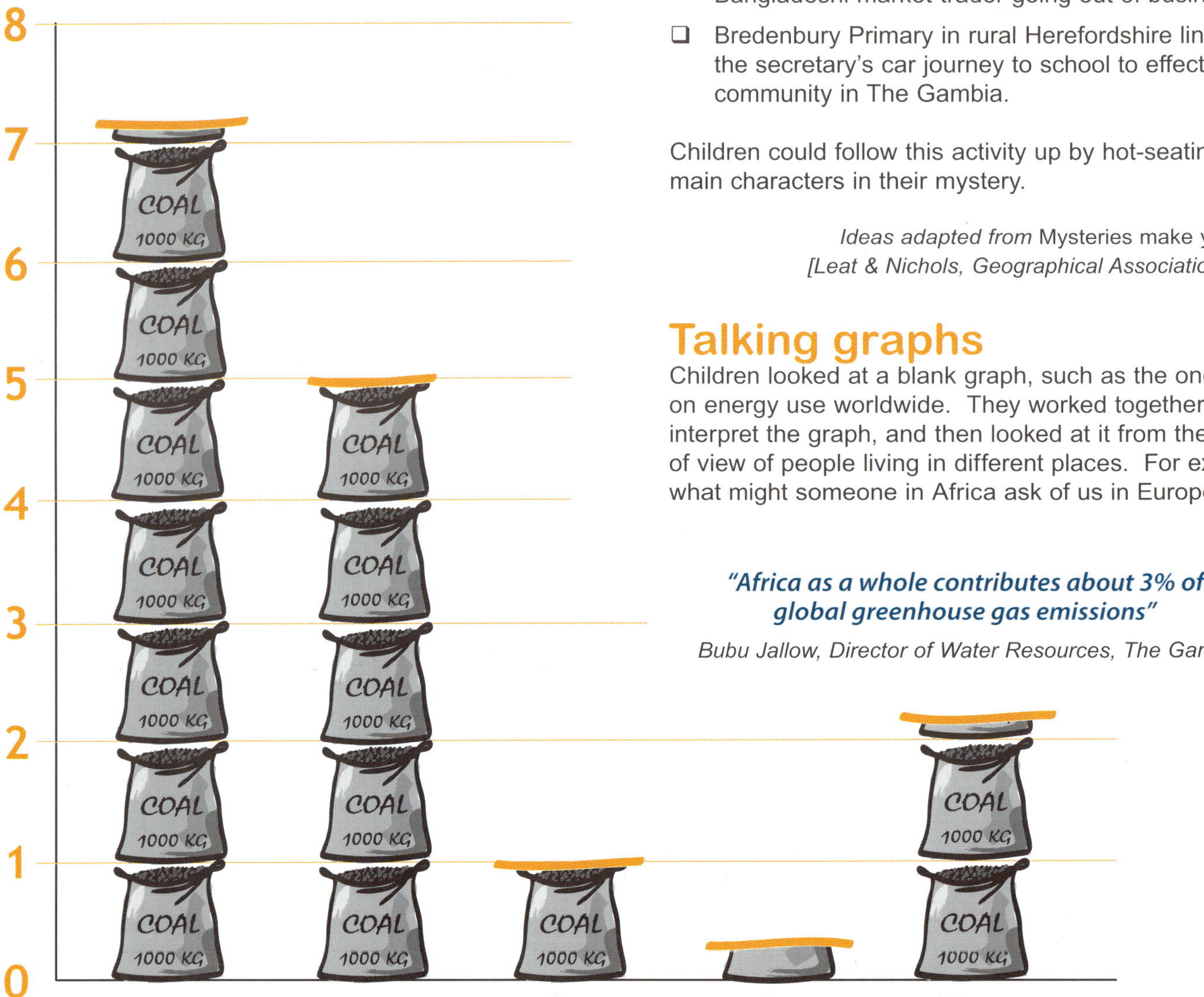

Global images: www.oxfam.org.uk/coolplanet; www.earthfromtheair.com; What is development? photopack [Tide~, 2003]

Why did Jade's mum have to buy a new carpet?

The River Severn flood burst its bank and water flowed into the houses by the river.

Jade and her family live in a lovely house down by the river.

When the snow in Wales melts, the water flows fast along the river into England and usually floods every year.

Flooding is more frequent as rain falls more often in downpours that run off the land, instead of as gentle rain that soaks in.

River banks have been built up in some places so that houses can be built on the river meadows, and so the flood waters have to flood somewhere else.

The River Severn is not dredged any more so it fills up with silt.

Global warming means that we have more extreme weather, - stormier, windier, wetter.

Jade's mum drives her to school by car over the main city bridge.

When the floods are very high, no one can cross the main bridge, and people have to go a long way round over the new bridge.

In the rush hour, the roads of Worcester are crammed with slow moving traffic.

Cars and other vehicles produce carbon dioxide.

Carbon dioxide is a greenhouse gas that adds to global warming.

Mystery A – Why did Jade's Mum have to buy a new carpet?

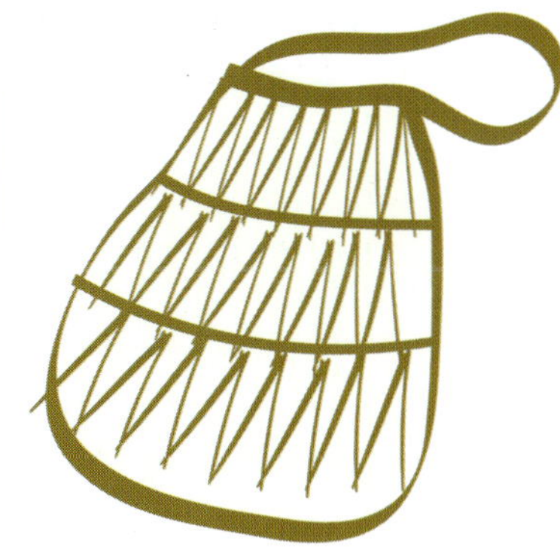

Mystery B – Why did Jaber's Mum have to use a string bag?

Cars produce carbon dioxide from the burning of petrol.

Dacca is a busy town with lots of traffic.

Why did Jaber's mum have to buy a string shopping bag?

Plastic shopping bags were banned.

Plastic shopping bags were found clogging up many drains during the recent floods.

The floods in Bangladesh are getting worse.

Trees in the Himalayan mountains are being cut down for firewood for people to cook on.

Where there are no trees to hold the soil into place, the heavy rains wash the topsoil down the mountains and the water spreads over the land.

Dacca is the capital of Bangladesh and is close to sea level.

Sea levels are rising as a result of global warming.

Carbon dioxide is a greenhouse gas that adds to global warming.

Global warming is believed to cause more extreme weather, so there will be more heavy rain and storms.

Science Across the World has a forum for children to share opinions on current topical scientific matters: see www.scienceacross.org

3 What can we do about it?

How can we teach about sustainable solutions to an issue that is not globally accepted? We have found plenty of available materials encouraging children to take action on climate change. Not all of these share debates about its nature and consequences.

Since uncertainty and complexity are very much part of what we were enquiring into, it was important to enable children to make links between the other key questions and ideas about effective action – and to come to conclusions about action for themselves.

We encouraged children to look at accepted wisdom in a questioning manner. Sometimes this meant the teacher playing Devil's Advocate [eg *"I do not want a wind farm to be built"*].

In our enquiry cycle action is not the end-point but stimulates further reflection, action etc. This is ongoing: learners never completely 'cover' the issue.

"Like many, I am still looking for answers to these questions. We need to keep asking questions ourselves and try to keep up to date with research."

Geography teacher, Wakeman School, Shropshire

Diary of an energy waster

- Working individually, children wrote an imaginary diary entry about a person who wastes energy in every aspect of their day.
- They passed this on to another child, who highlighted an area of waste.
- This was passed on to a third child, who suggested ideas for improvement.
- The class discussed what might happen if everyone behaved like the waster.
- We used this activity as a starting point for creative activities, assemblies, poetry, performance, D&T … [eg designing a waster's house].
- Some children rewrote their texts as a "diary of an environmentally friendly person".
- This activity could also lead into children undertaking a school energy audit.

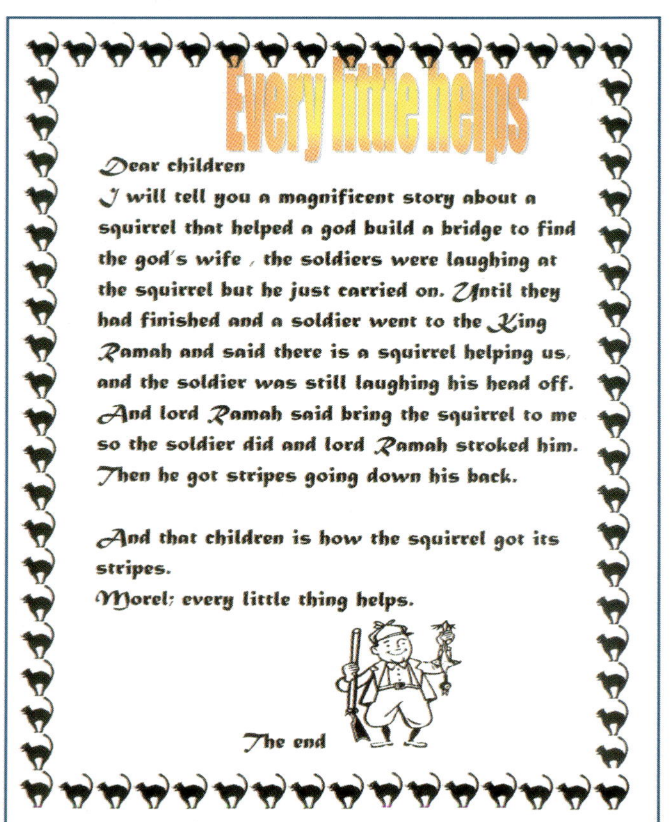

Every little helps

Dear children

I will tell you a magnificent story about a squirrel that helped a god build a bridge to find the god's wife , the soldiers were laughing at the squirrel but he just carried on. Until they had finished and a soldier went to the King Ramah and said there is a squirrel helping us, and the soldier was still laughing his head off. And lord Ramah said bring the squirrel to me so the soldier did and lord Ramah stroked him. Then he got stripes going down his back.

And that children is how the squirrel got its stripes.

Morel: every little thing helps.

The end

A huge global issue can seem overpowering. Children at Bredenbury Primary used an Indian traditional story, about a squirrel who helps to build a bridge, to explore the idea that 'every little helps'.

Low-impact living: www.theyellowhouse.org.uk www.cat.org.uk www.bio-diesel.org.uk
Carbon calculators: www.bestfootforward.com/carbonlife.htm www.futureforests.com http://coinet.org.uk/motivation/challenge/

The Mediterranean and the Big Chill

As part of a KS3 Geography enquiry at Wakeman School, Shrewsbury, children considered two potential future scenarios for Britain's climate:

- a Mediterranean scenario where warming continues unabated;
- a Big Chill – a mini ice age resulting from interruption of the Gulf stream by Arctic meltwater.

This proved a thought provoking activity for students which brought up many questions and concerns.

- Children investigated the evidence for each option, then chose which they would prefer to experience in 50 years time. Consideration was given to how it would affect aspects of lifestyle, jobs, leisure activities, food production, and disease.
- They imagined they were 50 years in the future, and considered how the climate would have changed.
- Finally, they wrote a letter back to someone in the present day telling them about the changes. This compared their lifestyles and suggested things that could be done to slow, minimise or mitigate the effects.

Energy audits

Most schools conducted energy audits, aimed at reducing CO_2 emissions. There are a number of useful schemes to support this [see websites below]. This included water use, which requires large amounts of electricity.

- Children completed energy and/or water-usage charts at home.
- They discussed energy/water conservation, and looked at presenting statistical evidence, as numeracy work.
- They surveyed the use of energy around school and noted areas of waste/overuse [heating, lighting and electrical appliances, water, recycling].
- The children surveyed other classes to identify recycling activity and made graphs/charts of the information collected.
- In some schools, children labelled all taps in school with water conservation stickers. In others, they labelled light switches. Some classes established energy monitors, whose job it was to ensure lights were switched off at break times.

Creative solutions

Children's own solutions

Food miles

Recycling, composting, water saving

Public transport, 'walking buses'

International agreements

Alternative energy: wind, wave, biofuels

Future inventions

New materials, fabrics etc

Energy saving devices

Land management – eg Dutch expertise used in The Gambia

Switch on your ideas

Children as researchers: www.woodland-trust.org.uk/phenology www.globe.org.uk www.climateprediction.net

Energy conservation: www.create.org.uk www.cse.org.uk www.est.org.uk

What can we do about it?

"Action may be about adaptation as well as mitigation. How can we adapt to climate changes as well as moderate them?"

Climate change: what is our future?

This used a full size version of the 'circles for change' image.

- On the left children placed ideas about what they currently do, and what they thought was happening at a wider scale.
- They conducted audit trails around the school, looking at energy use and waste, and brainstormed ideas to inform whole school actions.
- They also considered ways in which their school and community might adapt to different climate change scenarios.
- On the right they placed ideas about what they hoped would happen in future.

Shaking hands

At the environmental event, *Green and Kickin',* in Wolverhampton, children drew around their hands and on each finger listed one activity which they would undertake. This was linked to a second hand, on which they listed actions for others to undertake, eg school council, government, family, neighbours. It was important to give children a chance for reflection before making their decisions.

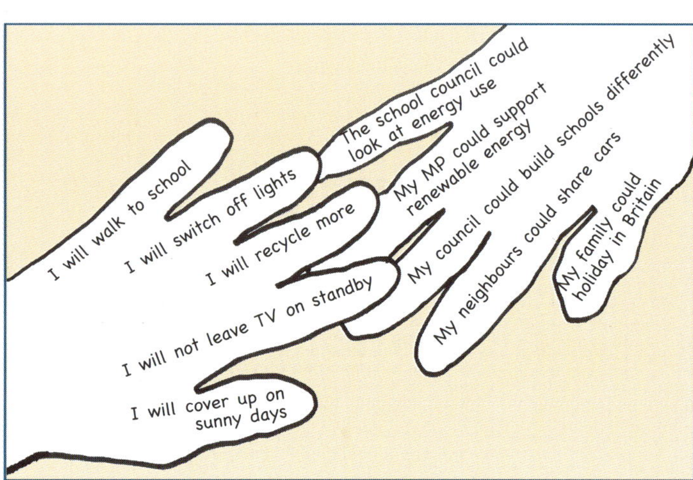

Circles for change

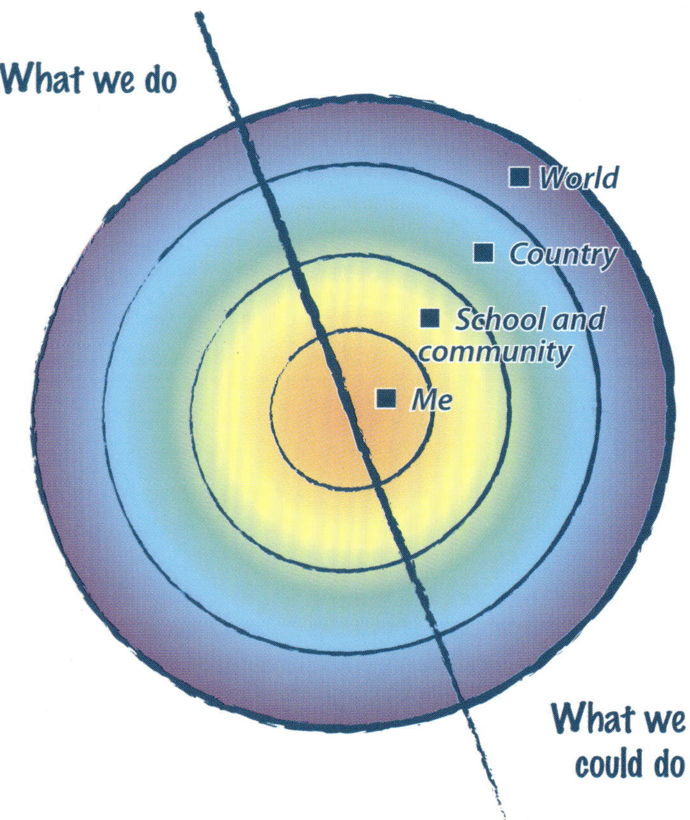

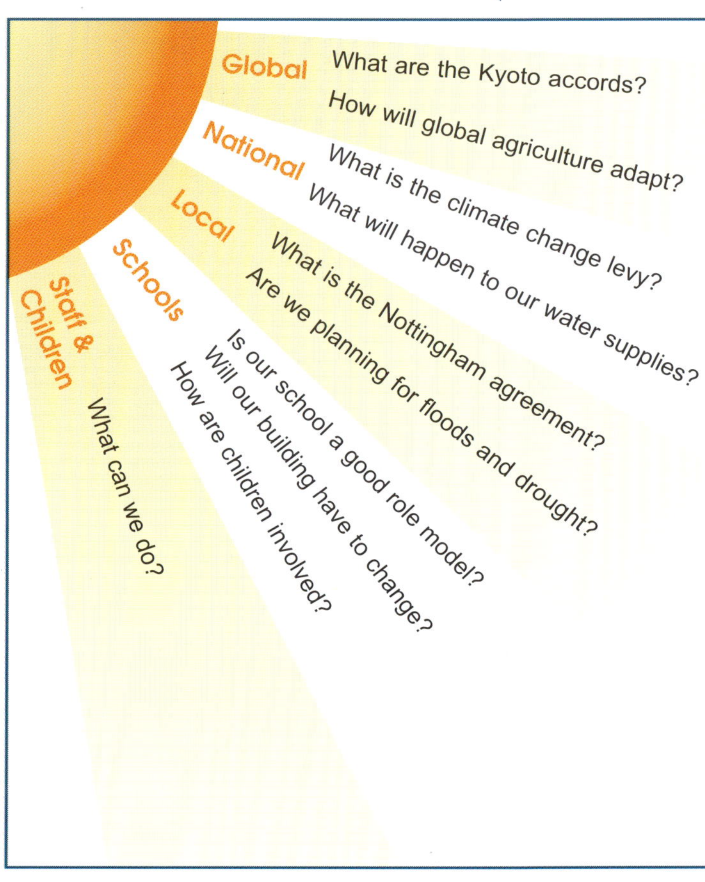

An extended version of these prompts can be found on www.tidec.org

Action and campaigning: Carbon Trust [businesses] www.thecarbontrust.co.uk
Climate Action Network [international] www.climatenetwork.org
Nottingham Declaration: www.devon.gov.uk/cc-signnotdecl.pdf

Sharing our learning, influencing others

Year 4 children at Somerville Primary in Birmingham worked with Severn Trent Water's Barston Education Centre to investigate climate change. The children reviewed their learning and findings, and prepared a presentation for use at a climate change day at the Centre. This day concluded their project, and focused on developing ideas about how to protect the environment.

- The children felt empowered to raise awareness with a wider audience.
- They produced a mind map of the groups they wanted to target and the methods they might use [below].
- They carried out some of these activities and sought feedback.

Other schools also made presentations:

- Children at Kingswinford Secondary and St Mary's CE Primary School in Dudley shared ideas about effective action at a 'marketplace' event.
- At Foundry Primary School, Birmingham, children made a presentation to parents and local residents.
- At Bredenbury Primary School, Herefordshire, and Albert Bradbeer Junior School, Birmingham, children worked on persuasive text, and produced booklets sharing their thoughts with parents and peers. They considered where else they might go [eg doctor's waiting room, library].
- Children's ideas at Holbrook Primary, Coventry, were shared with parents, and their work exhibited as part of a dedicated Science week.

Other ways of sharing ideas included posters and PowerPoint presentations.

You could also assign groups to a local business to monitor sales of 'green' products [eg low energy lightbulbs] and advise managers on promoting them.

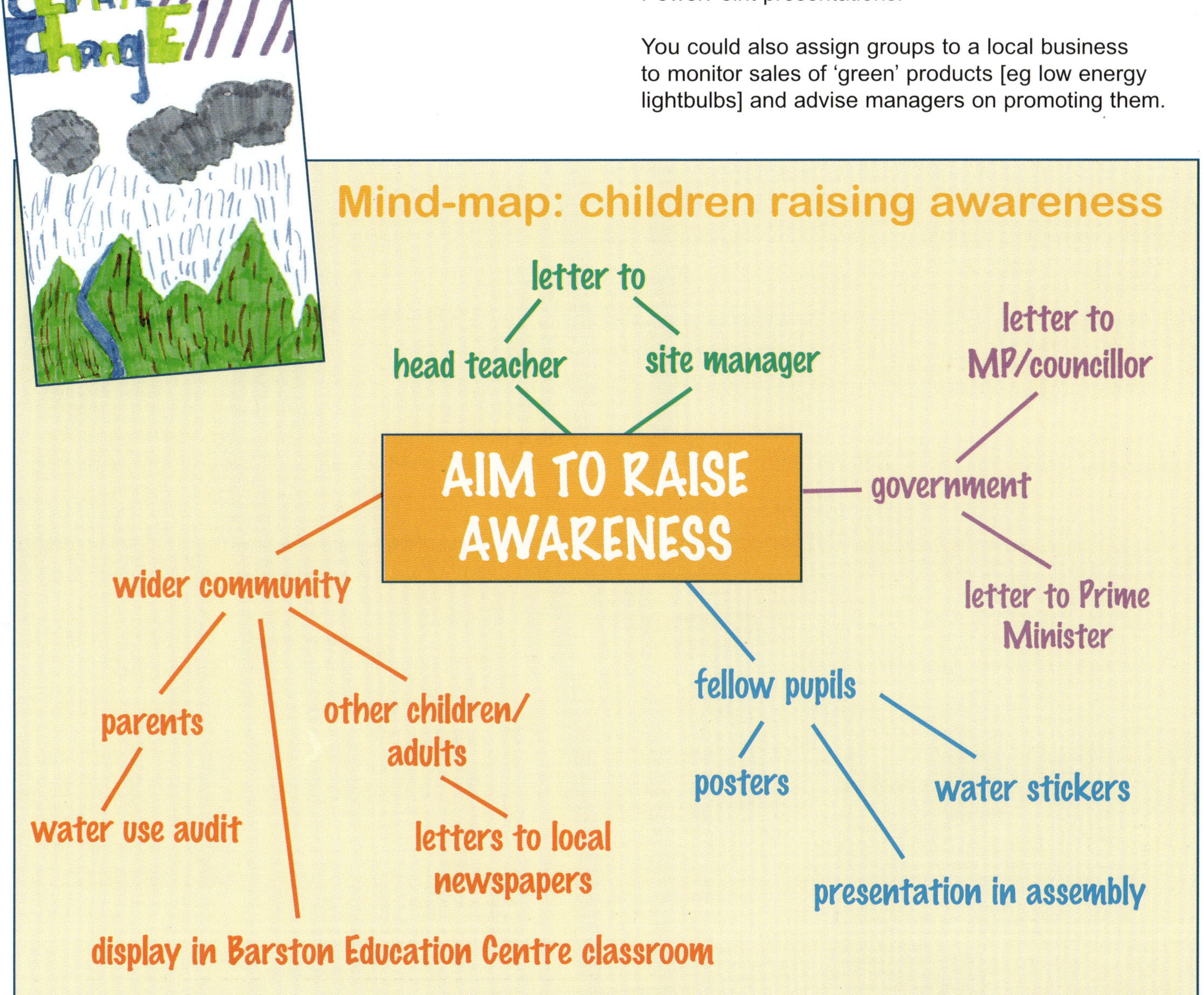

World Wide Fund for Nature www.panda.org/climate
Rising Tide www.risingtide.org.uk
Friends of the Earth www.foe.co.uk/campaigns/climate

4 What have we learned and how?

These are crucial questions in the enquiry learning process. We found that it helped if children were engaged, throughout their enquiry, in a process of recording and reflecting on their learning. Some of the activities in this section therefore draw on activities which were started right at the beginning of the enquiry process.

How did we find out information?

What resources did we use?

What help did we need?

Who helped us?

How did we decide what we wanted to know?

How did we share what we knew?

How did we make our ideas interesting to others?

What do you know? What do you want to know?

Children noted what they ***K****new* at the start of the topic. They then identified gaps in their knowledge and what they ***W****anted* to know.

During their enquiry, they returned to the grid and noted what they had ***L****earnt* and ***H****ow* [eg what were their sources of information? what ways of finding out did they find most useful?]

Know	Want to know	Learnt	How
That climate change affects everyone	What sort of affects it might have	It does not affect everyone the same way: eg, it can lead to floods in some places and drought in others.	From books and the internet, by talking with other children in the class.

"It's amazing what children can do when they put their mind to it!"

KS2 teacher, St Mary's CE Primary, Kingswinford

The Enquiry Learning Cycle

The four key questions were placed on a big circle on a notice board in the classroom. As the enquiry progressed, children added ideas and further questions to the display, as a visual map of the process.

The process didn't always follow the number sequence [and this was especially the case with question 4 itself].

1 What is climate change?

2 Why does it matter?

3 What can we do about it?

4 What have we learned?

In Bredenbury Primary School, children presented their ideas in a booklet which used the four key questions as section headings.

Mind mapping online - www.mayomi.com   Search about mind-mapping at www.bgfl.org

Mind maps

We used mind maps to guide children's learning throughout, and revisited these periodically. Some of us used mind-mapping software to do this.

Margolis wheel

This technique was used as an end-point to enquiry, but it could be brought in at other stages in the process. The wheel is a bit like hot-seating, in that some children are the questioners and others in role as the experts being questioned.

- ❑ The experts [half the class] formed an inner wheel.
- ❑ Questioners formed an outer wheel, and were given two minutes to question and listen.
- ❑ They then moved on one place, and heard the answers from the next expert.
- ❑ After a while, questioners and experts switched roles.

- revisit the mind map / Have we found out what we wanted to know?

How did we learn best?
- internet
- books
- visits/visitors
- photos/video
- drama/role play
- talking to family

How can we share what we've learned?
- presentations
- displays
- posters
- video/drama

What had most impact?
- statistics/figures
- human stories

How do we avoid trivialising the issues?

WHAT HAVE WE LEARNED? - AND HOW?

Part of a mind map which used computer software

Learning priorities

Children from Foundry Primary School in Birmingham worked with Centre of the Earth.

- ❑ Children were asked what they had learned … and how they had learnt it.
- ❑ Their responses were recorded on a flip chart.
- ❑ Each child had one vote each for the most effective, most useful and most memorable item on the flip chart.
- ❑ The four or five most popular responses were then recorded.
- ❑ This helped with future enquiries.

Margolis wheel

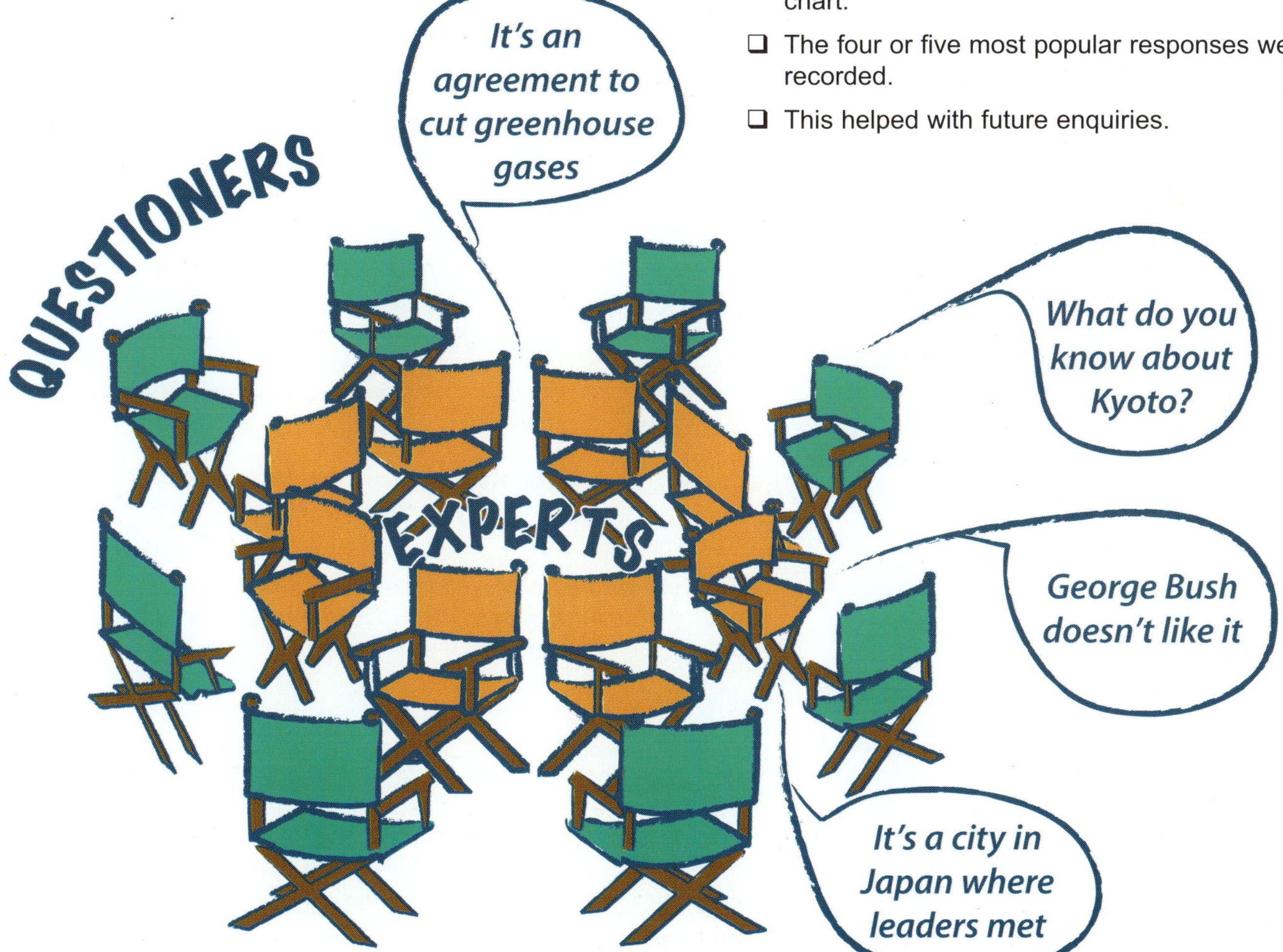

For links to presentations by participating schools see www.tidec.org

What have we learned and how?

Sharing our learning

Year 4 children, Somerville Primary School, Birmingham

Following their enquiry, children made a presentation for the wider community. This shared their ideas, and the ways in which information had been gathered. As a culmination of the whole project, the children each made a pledge or a statement, reflecting their feelings about the issues they had researched. These were linked together to make a chain, which formed a visual reminder.

The teacher and children identified a number of learning outcomes, including:

- ❑ increased use of speaking and listening skills;
- ❑ collaborative approach, managing groups;
- ❑ strengthened link between the school and local secondary schools;
- ❑ Increased use of e-learning and enhanced ICT skills;
- ❑ cross-curricular links.

They also identified gaps. For example, did children have a good enough understanding of specific question terminology by the end of the enquiry?

Year 6 children, Foundry Primary, Birmingham

As a culmination of work on climate change and energy issues, children made a presentation at an after-school event attended by a hundred parents, plus other local residents and organisations.

The presentation began with the children outlining some statistics about the world as if it was a village of 100 people. They explained some of the implications of global warming for the most vulnerable members of this village.

They prioritised the inequities resulting from global warming, not least how many 'developing' countries contribute little to carbon dioxide emissions but are particularly vulnerable to the consequences of climate change.

They performed the poem *Beware* about the effects of sea level rise. This poem was written by Grade 6 students at Bakau Newtown Primary School, The Gambia, and had particular significance for the class because one of the students had family links there. They also presented findings from a home energy survey, with suggestions about how we can all help save energy.

"For our global village global warming means:

- ❑ *food shortages*
- ❑ *water shortages*
- ❑ *more refugees*
- ❑ *more disease"*

"In August 2003, 35,000 Europeans died due to a heat wave which swept Europe"

"Global warming is melting ice caps and glaciers which puts millions of people at risk from flooding"

"The world is warming faster than at any time in the past 10,000 years"

"20% of families surveyed leave the TV on standby instead of turning it off. This is a big energy waster. Remember, turn off the TV when you have finished"

BEWARE

"The millions you throw aren't you aware
I work night and day and never go to sleep
While you snore aren't you aware
Of me the ocean you should beware
For I continue to work while you sleep
Beware beware you better beware."

Other schools made presentations as part of their work [see page 25]. In order to do this, children needed to take stock of their learning, and think about the ways of presenting ideas which worked best for them; this proved a good way of integrating thinking about learning with an activity which had purpose and meaning for the children.

The world as a village ... of 100 people:
www.mysterra.org/webmag/coup-de-coeur_en.html

Evaluating *how?*

Children used the following prompts as the basis of an oral evaluation [eg at circle time] and as the basis of a display [ie as an inner circle, with their comments around the outside].

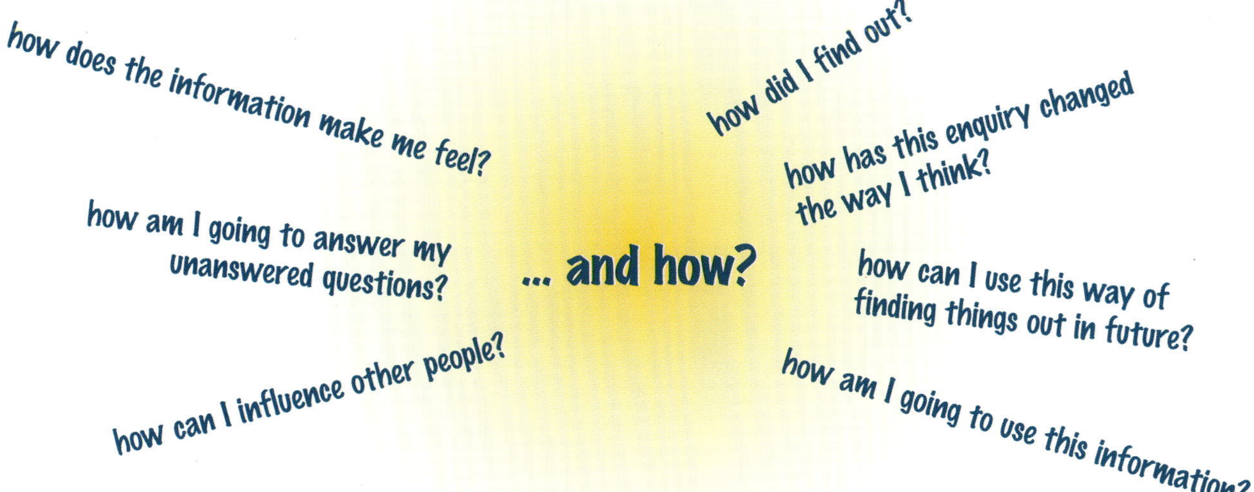

Evaluating sources

Children used this framework to record and evaluate their sources of information.

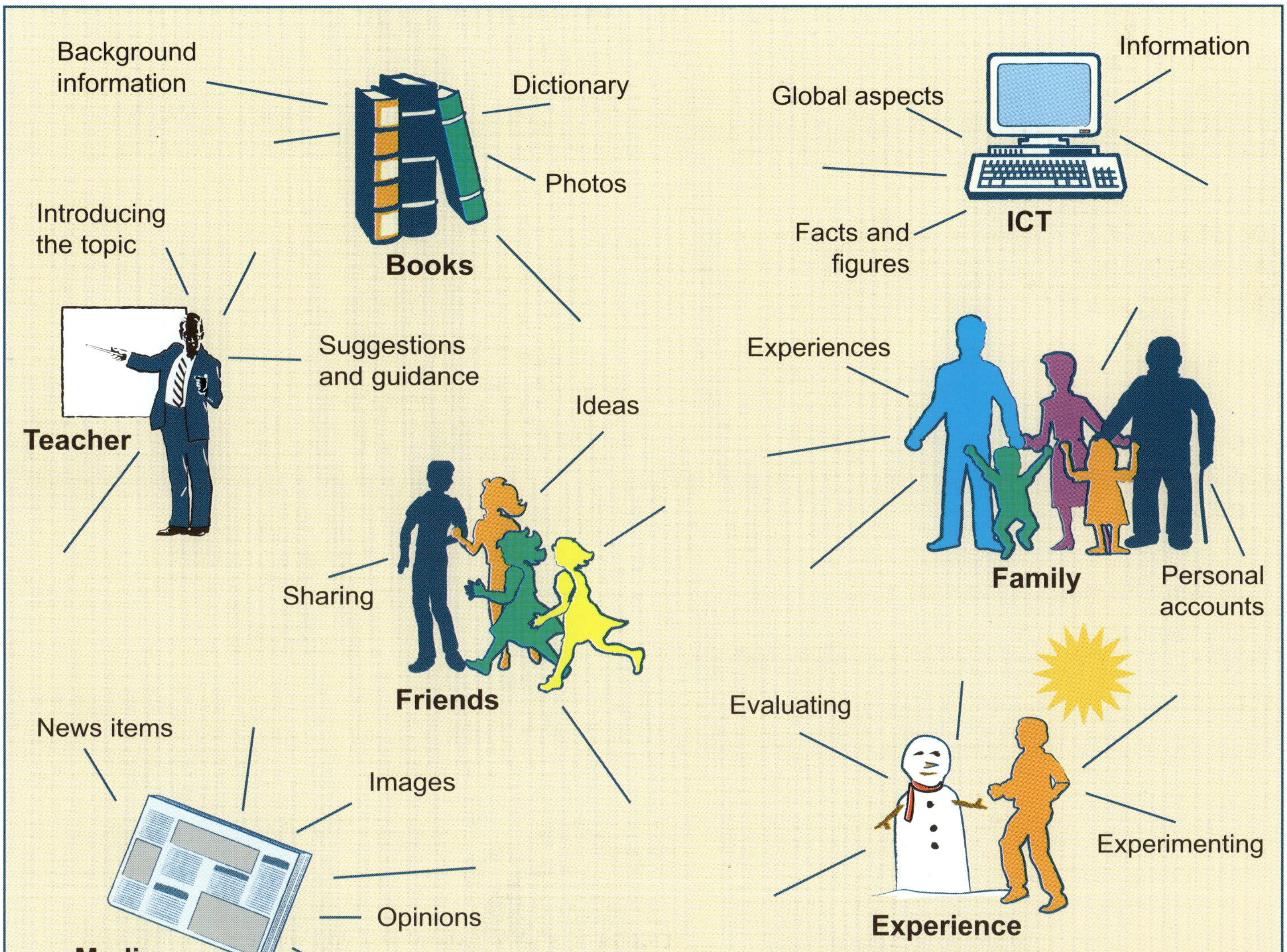

Background information
Dictionary
Photos
Books

Global aspects
Information
Facts and figures
ICT

Introducing the topic
Suggestions and guidance
Teacher

Experiences
Family
Personal accounts

Ideas
Sharing
Friends

News items
Images
Opinions
Media

Evaluating
Experimenting
Experience

This framework can be downloaded as a PDF

... of 1000 people www.angelfire.com/il/adventureclub/village.html

Also in the series **Sustainable development, local and global:**

Learning today with tomorrow in mind – *sustainable development education* [2000]

Food and farming, local & global – *planting ideas, growing ideas* [2004]

Water issues, local & global – *exploring sustainable development at KS2* [2005]

If you are interested in local and global sustainable development, the following titles from Tide~ may also be useful:

Waking up – *using plants to explore sustainable development at KS2* [with Birmingham Botanical Gardens Study Centre, 2000]

Educating for sustainability – *The Gambia and the UK* [with the National Environment Agency, The Gambia, 2002]

Lessons in sustainability – *What on earth is happening? How do we respond?* [2003]

The development compass rose – *a consultation pack* [1995]

Resources for teachers

Climate change poster New Internationalist

Teaching about climate change – *Cool schools tackle global warming* Tim Grant and Gail Littlejohn, New Society Publishers, 2001

Climate chaos – *information for teachers* WWF-UK, 2005 [downloadable online from www.wwflearning.co.uk]

Shout about climate change – *Education pack for Key Stage 3* Friends of the Earth, 2005 [downloadable online at www.foe.co.uk/learning/educators/shout_about]

Resources for students

Climate change – *Our impact on the planet* Simon Scoones, Hodder Wayland, 2001.

Global warming – *The threat of Earth's changing climate* Laurence Pringle, SeaStar Books, 2001

Energy Matters pack CREATE/CSE

Energy forever? series: *Fossil fuels, Geothermal & Bio-Energy, Nuclear Power, Solar Power, Water Power, Wind Power.* Ian Graham, Hodder Wayland, 1998

Sustainable world: Energy Rob Bowden, Hodder Wayland, 2003

Background resources

Funny weather we're having at the moment, isn't it dear? Kate Evans, Rising Tide, 2003

High Tide – *News from a warming world* Mark Lynas, Flamingo, 2004

The No-nonsense guide to climate change Dinyar Godrej, New Internationalist, 2001

NorthSouthEastWest – *a 360° view of climate change* The Climate Group, 2005

All of these resources are available from The Tide~ Centre

Online resources

Websites which we have found useful are listed throughout the text. We also draw your attention to the following.

Teaching resources

Atmosphere, Climate & Environment: www.ace.mmu.ac.uk

Global Eye: www.globaleye.org.uk

Global Express [Edition 23]: www.dep.org.uk/globalexpress

Global warming enquiry [written by Stuart Purves, Moorside High, Stoke-on-Trent]: www.sln.org.uk/geography/enquiry

Background

Climate change projections: www.UKCIP.org.uk/scenarios

Exxonmobil [oil company] – search on 'climate change': www.exxonmobil.com

The Guardian [newspaper]: http://politics.guardian.co.uk/green

Hadley Centre [research]: www.met-office.gov.uk/research/hadleycentre

Mark Lynas [author and journalist]: www.marklynas.org

Real Climate [scientific debate]: www.realclimate.org

Tyndall Centre [research]: www.tyndall.ac.uk

WMnet Climate change in the curriculum programme
Climate change portal:
http://climatechange.wmnet.org.uk

To order Tide~ resources online, download some of the stimulus materials in this resource, and find out more about professional opportunities for teachers on this theme, see the Tide~ website:

www.tidec.org

Extended acknowledgements

These materials arise from the creativity of very many teachers across the West Midlands region.

The main writing group and others playing a lead role are acknowledged on the inside front cover.

Five local working groups were involved in the main creative classroom work leading up to this publication:

Birmingham

Facilitators: Malcolm Smith, *Barston Education Centre;* Sonia Wright, *Bell Heath Study Centre;* Andrew Simons, *Centre of the Earth.*

Working group: Gillian Beck, *Boldmere Junior;* Anna Jaremko, *Boldmere Junior/New Hall JI;* Lesley Geoghegan, *Somerville JI.*

Also involved: Sue Penhallow, *Albert Bradbeer Junior;* Paul Archer, *BASS;* Terry Pugh, *Bockleton Study Centre;* Kirk Wellington and Deborah Latham, *Foundry JI School;* Ian Yates, *James Watt Junior.*

Dudley and Wolverhampton

Facilitators: Steve Lockwood, *Dudley LEA;* Sue Shanks, *Natural Curriculum Project, Wolverhampton.*

Working group: Kerry Harris and Adrian Hyde, *Blanford Mere Primary;* Jodie Mills, *Highfields School, Wolverhampton;* Lisa Parkes, *St Mary's Primary, Dudley [now at Alder Coppice Primary, Dudley];* Nicki Spilman, *Kingswinford School, Dudley;* Denise Ward, *Kingswinford School.*

Also involved: Greg Jones, *Black Country School Improvement Partnership;* Alyson Whittaker, *Bramford Primary.*

Coventry

Facilitators: Sarah Oakley, *Agenda 21;* Phil Leivers, *Coventry LEA.*

Working group: Simon Bonney, *Alice Stevens Special School;* Majella Forrester, *Foxford School;* Lisa Ambler, *GLOBE;* Jane Bufton, *Holbrook Primary;* Sarah Jackson, *Performing Arts Service;* Lynn Melling, *WEEAC.*

Also involved: Mr N Owen, *Allesley Primary;* Elaine Gough, *Bishop Ullathorne School;* K Bonehill, *Clifford Bridge Primary;* Bill Johnson, *Coventry City Council;* Jane Barker, *Coventry LEA;* Veronica Parrell, *Holbrook Primary;* Elizabeth Chester, *Manor Park Primary;* Karen Hawcutt, *Radford Primary;* Jackie Dines, *Sidney Stringer CTC;* Deborah Knott, *Warwickshire Wildlife Trust [now at Birmingham Botanical Gardens];* Rose Smith, *Warwickshire Education Ranger;* Rohini Corfield, *Warwickshire LEA.*